Chasing Rainbows with Nashe

A Science Story About Light

by Rudo Kashiri

Curious Minds Explorers Publishing

A Curious Minds Explorers picture book

Chasing Rainbows with Nashe

A Science Story About Light

Illustration by Roshanay
Published by Curious Minds Explorers Publishing
Stories That Launch Curiosity
www.curiousmindsexplorerspublishing.com

ISBN: 979-8-9996203-4-7

Printed in the United States of America

This book is a work of fiction inspired by real science.
Names, characters, places, and events are used to encourage curiosity and learning and are presented in a fictionalized format.

This book is dedicated to Thelma.

You are a woman of extraordinary courage.

In the face of adversity, you persevere and emerge with renewed strength, much like a rainbow that appears after a storm. I am profoundly proud of you.

"Hope is being able to see that there is light despite all of the darkness."

— Desmond Tutu

This book belongs to

____

Welcome to Nashe's Rainbow Adventure

Have you ever spotted a rainbow and wondered, "Wow! How did that get there?"

Nashe has questions like that, too, and she doesn't just stare at the sky. She's ready to chase the rainbow, test ideas, and uncover its secrets.

With her wise grandmother, Ambuya, by her side, Nashe sets off on a colorful quest. Together they'll splash in puddles, chase warm sunlight, and discover how water and light, with a dash of wonder, team up to paint rainbows across the sky.

So grab your rain boots and your biggest questions.

Remember:

Curiosity is your superpower.

Every "why?" is the start of a brand-new science adventure.

Rain *tapped, tap, tapped* against the window as Nashe stretched out on the couch, her chin resting in her hands.

Outside, the world looked gray and drippy. Puddles swelled on the sidewalk, and thick clouds blanketed the sky.

Her favorite book lay closed beside her, its red cover catching the dim light from the rain-streaked window.

She let out a big, dramatic sigh. "Rainy days are always so boring," Nashe grumbled, watching the raindrops slide down the glass.

A few streaks raced, zigzagging and leaving twisty trails behind them. Nashe followed the racers with her eyes, but the room still felt slow and sleepy.

As Nashe stared at the clouds, a tiny spark flickered in her chest. Maybe something interesting was hiding behind all that gray.

She scooted closer to the window and peered out, as if looking hard enough could reveal a secret.

Somewhere out there, behind the rain, the sun was still shining. Nashe pressed her fingertip to the cool glass and whispered, "If light can find a way through, maybe I can, too."

Later, the rain softened to a gentle whisper against the window. A warm sliver of sunlight slipped through the gray clouds, and the whole room seemed to glow.

At once, Nashe hurried to the glass and pressed her nose close, her breath fogging a little circle on the window.

Her eyes grew wide. "Ambuya, look! Look!" she shouted, her voice bright as sunshine. "There's a rainbow! It's so bright!"

She bounced on her toes, hardly able to stand still. "I've never seen one this close before! Do you see all the colors: red, orange, yellow, and more?"

She traced the arc in the air with her finger. "It's like someone painted the sky!"

Ambuya came to stand beside her, smiling warmly. "I see it, Nashe. Isn't it beautiful? Rainy days can bring the best surprises."

Ambuya's eyes twinkled as she looked from the rainbow to Nashe. "Do you know what makes a rainbow appear after the rain?" she asked.

Nashe shook her head, hope bubbling in her voice. "Is it... magic?"

Ambuya chuckled softly. "It can feel like magic," she said, "but it's science at work. Rainbows happen when sunlight meets tiny drops of water floating in the air."

"In the stories I grew up with, rainbows are messages from the sky, little reminders to look closely at the world around us. And there's science, too," she added. "When sunlight meets raindrops, something amazing happens."

Nashe leaned in, eyes wide. "So what does the sunlight do?"

Ambuya traced a rainbow shape in the air with her finger. "When sunlight enters a raindrop, it bends and spreads into colors. That bending is called refraction."

Nashe's mouth fell open. "So the colors were hiding inside sunlight the whole time?"

"That's right," Ambuya said with a nod. "Sunlight looks white, but it's made of many colors mixed together. Raindrops help separate those colors so our eyes can see each one."

Nashe grinned, excitement bubbling up again. "So if it rains and the *sun* comes out... there could be a rainbow?"

"Exactly," Ambuya said. "All you need is sunlight, raindrops, and a little curiosity." Her smile shone almost as bright as the rainbow.

Nashe looked back at the rainbow, then at Ambuya. "So... what is light?"

Ambuya's smile softened. "That's a wonderful question, Nashe," she said. "Light is a kind of *energy*, like a super-fast runner from the sun. It zooms through space and still takes about *eight minutes* to reach us!"

"So light is always moving even when we can't see the beam?" Nashe asked.

"That's right," Ambuya replied. "Light is always traveling. It can bounce off mirrors and shiny things, bend through water or glass, and spread into colors."

Nashe tilted her head. "Did people always know what light was?"

Ambuya shook her head gently. "No, Nashe. Long ago, people learned by watching the world closely and noticing patterns. In our culture, paying attention is the first step to learning."

Nashe nodded, her curiosity growing. "So every time I see something, it's because light is bringing *information* to my eyes?"

"Exactly!" Ambuya said warmly. "Without light, our world would be dark. All the colors we love would disappear."

Nashe grinned, her eyes shining. "I never knew light could do so much. It really is a secret superpower!"

Ambuya nodded. "It is. And the more you learn about light, the more amazing the world becomes."

Nashe looked at the sunbeam one more time. "So *light is energy,*" she said slowly. "And it's how my eyes can see the rainbow."

Ambuya nodded. "Yes, my clever girl. Light carries information to your eyes, colors, shapes, everything."

Nashe's face brightened. "Then, can we use that energy to *make* a rainbow on purpose?"

"We can," Ambuya said. "Let's go outside and try."

Nashe grabbed her rain boots and hurried after her. The air smelled clean and fresh after the rain. Sunlight peeked through the clouds.

"Ready for a science adventure?" Ambuya asked. "Let's do a quick *experiment* to learn more about rainbows."

"I'm ready!" Nashe said, practically bouncing. "Can I try?"

Ambuya handed her a garden hose. "One important thing," she reminded gently. "Keep the sun behind you."

Nashe nodded and gripped the hose with both hands. She aimed the spray up into the air.

"If I make the water into tiny drops like rain, maybe I'll see a rainbow!" she guessed, glancing at Ambuya.

Ambuya smiled. "Let's watch what happens when sunlight and water meet."

A sparkling mist floated in front of them. Nashe held her breath, waiting for something amazing to appear.

"Do you think it'll work?" she whispered.

Ambuya leaned in. "What do you *predict*, Scientist Nashe?"

Nashe narrowed her eyes in concentration and aimed the hose right into the sunlight. "I think if the sunlight hits the water just right, I'll see a rainbow!"

"Woohoo!" Nashe cheered, doing a quick little spin. "Ambuya, look! The rainbow is right there in the spray! I can see red, orange, yellow, green, blue, and violet."

With a small twist of her wrists, Nashe nudged the hose a little to the left. The rainbow slid with it, shimmering in the mist. "If I move the water, the rainbow moves, too. It's like I'm painting with sunlight!"

Ambuya watched from a few steps away, her smile gentle and proud. "You're noticing so much, Nashe. Why do you think the colors always show up in the same *order*?"

Nashe thought for a moment, then grinned. "Is it because each color of light bends a little differently in the tiny water drops, so they always line up the same way?"

"That's right," Ambuya said. "You're testing your ideas just like a scientist. What do you predict will happen if a cloud covers the *sun*?"

Nashe glanced up at the sky. "Maybe the rainbow will disappear, because without sunlight, the water can't bend the light into colors."

Just then, a cloud drifted over the *sun*. The rainbow in the mist grew faint, then vanished.

When the sunlight peeked out again, the colors popped right back into place.

She turned back to the sparkling spray and watched the colors wiggle and dance. "I hoped for a rainbow... and there it is!"

Next, Ambuya handed Nashe a bottle of bubble solution. "Now let's see if bubbles can make rainbow colors, too."

Nashe dipped her wand and blew a slow, careful breath. Bubbles bobbed into the sunlight. "Ambuya, look! Every bubble has a rainbow swirling on it!"

Laughing, Nashe chased them as they drifted and *popped.*

Then Nashe paused, watching the colors slide across a bubble's skin. "Why do they change when the bubble moves? Is a bubble rainbow the same as a sky rainbow?"

Ambuya stood beside her. "You're noticing patterns, Nashe. When you tilt the wand, what happens to the light?"

Nashe squinted at a bubble. "Maybe light hits the bubble and starts bouncing back and forth," she said.

Ambuya nodded. "A bubble's thin surface makes light reflect and mix in ways that create shimmering colors."

She pointed toward the sunbeam. "Try blowing bubbles in the shade, then in the sunlight. What changes?"

Nashe blew again. Some bubbles drifted into the shade, and their rainbow shine faded. Others floated through the sunlight and glowed.

"The rainbow disappears when a bubble pops," Nashe said, "but it comes back on new ones!"

Ambuya smiled proudly. "Every question helps you learn something new."

After the last bubble *pop*, Ambuya and Nashe headed back inside.

"One more experiment," Ambuya said, "with a tool scientists love."

Nashe's eyes lit up. "A real science tool? Yes!"

Ambuya gently set a prism on the table. "Watch closely, Nashe," she said.

"Light bends as it passes through the prism and spreads into colors called a spectrum."

Sunlight streamed through the prism. A rainbow spilled across the table, bright bands lined up side by side.

Nashe's eyes widened. "It's just like the rainbow in the sky, but smaller! Is this how scientists learn about light?"

"Yes," Ambuya said. "They shine different kinds of light through tools like this and *compare* the colors. The color pattern gives clues."

Nashe traced the rainbow with her finger, careful not to block the light. "So every rainbow is sunlight bending and spreading into colors?"

"That's right," Ambuya replied. "And now you know a new science word: *spectrum*."

Ambuya leaned closer. "This is like what scientists do with a tool called a *spectroscope*," she said. "It spreads starlight into a rainbow strip."

"In that spectrum strip, scientists can see thin dark lines that act like clues."

Nashe smiled. "Like secret codes?"

"Yes," Ambuya said. "Those lines are clues that tell what the star is made of."

Nashe stared at the rainbow, and her brain filled up with questions, *plop, plop, plop*, like raindrops.

"Ambuya," she said, "what's on the other side of a rainbow? A pot of gold?"

Ambuya smiled and drew a gentle rainbow curve in the air. "No pot of gold, my dear. A rainbow doesn't really have an *end*."

Ambuya pointed toward the ground. "A rainbow is actually a *full circle* of colors. From the ground, the horizon blocks part of it, so we usually only see the top, like a smile in the sky."

Nashe squeezed her eyes shut and pictured herself way up high, above trees and rooftops, peeking down at a glowing loop of color.

"So if I were up in the sky, I could see the *whole* rainbow?"

"That's right," Ambuya said. "Pilots and mountain climbers sometimes spot a full-circle rainbow, like a colorful ring. It depends on where *you* are and where the sun is shining."

Nashe grinned, imagining herself floating above the clouds, surrounded by rainbows like hula hoops. "Ohhh, I want to see a whole rainbow someday!"

Ambuya nodded. "Keep wondering, Nashe. Curiosity helps you notice new things even in the same old sky."

Later that day, the rainbow faded into the evening light. Its colors melted into the golden glow of sunset. Nashe watched, quiet and calm.

"Even when the rainbow disappears," she whispered, "my questions stay."

Looking up at Ambuya, Nashe asked, "Does it ever stop? All the questions?"

Ambuya squeezed her hand. "Never," she said. "Every scientist begins with a question, and you've already started."

Nashe stood a little taller. "I'm not scared of not knowing," she said. "That's where the adventure begins."

Ambuya nodded. "Wonder is the beginning of wisdom, my child. Keep your eyes open and your heart brave, and you'll find your own rainbows again and again."

Nashe tipped her face to the sky, feeling hope and curiosity swirl inside her. "I'll keep wondering, Ambuya. I promise."

Ambuya smiled and gave Nashe's hand a gentle squeeze. "Good," she said. "Keep those questions close. They'll take you far."

With a grin, Nashe said, "Good, because I've got a lot of *whys*."

Ambuya laughed softly. "Then tomorrow we'll chase one of those *whys* together," she said.

Nashe took one last look at the fading sky. Somewhere behind the clouds, light was traveling, bending, waiting to be discovered. She was ready for the next question, too.

Ambuya's Science Corner

Long before lab coats and science books, people learned by watching the world around them.

In many African cultures, elders taught children to observe the seasons, ask questions, share stories, and learn by trying.

Today, scientists do many of the same things. They observe nature, test ideas, and use tools like prisms and spectroscopes, always asking, "Why?"

When scientists spread starlight into a spectrum, the patterns of lines act like clues, almost like fingerprints.

Science grows when we stay curious and pay attention to small details.

Light from a star enters a spectroscope and spreads into a spectrum.

Did You Know?

The sunlight you see now left the Sun about eight minutes ago and traveled about 93 million miles to reach you.

Try It: Make a Rainbow on Paper

Ready for one more rainbow *investigation*? Make a rainbow indoors using water and light.

Materials

- A clear glass or jar
- Water
- A sunny window *(or a flashlight)*
- A white sheet of paper

Grown-up note: Don't let kids look directly at the Sun. Don't shine the flashlight into anyone's eyes. Do this over a towel in case of spills.

Steps

1. Fill the glass about **3/4** full of water.
2. Set it on a sunny windowsill. *(In a dim room, have a grown-up shine a flashlight through the side of the glass toward the paper.)*
3. Hold the paper on the other side of the glass (where the light comes out). You can also lay it on the table or prop it up. Move it slowly until you see a rainbow.
4. Tilt the glass a tiny bit (over a towel) and try a few angles.
5. Record it: draw what you see.

Wonder Questions

- What changes when you tilt the glass?
- How can you make the rainbow bigger or brighter, and do the colors stay in the same order?

Think like a scientist: Keep the glass, water level, and paper the same, then change only one thing (angle or distance) and notice what happens.

Try this next: Add one drop of food coloring, or try a different container. Does the rainbow change?

Author's Note

As a science educator, I've seen how quickly children turn everyday moments into big questions.

In many African traditions, knowledge is shared across generations through careful observation, listening, and storytelling. Those ways of learning have shaped how I think about science, and they inspired Ambuya's role in this story.

Chasing Rainbows with Nashe brings these ideas together through story, questions, and discovery.

Rainbows invite wonder, and they also follow clear scientific rules. Through conversation and hands-on experiments, children can begin to understand not only how the world works, but also how knowledge itself grows.

My hope is that this book encourages families to look up together, ask questions together, and discover that science isn't separate from culture or everyday life, it's woven through all of it.

Rudo Kashiri

Curious Minds Explorers Publishing

www.ingramcontent.com/pod-product-compliance
Lightning Source LLC
LaVergne TN
LVHW070900210926
848482LV00016B/981
* 9 7 9 8 9 9 9 6 2 0 3 4 7 *